# YELLOWSTONE

## THROUGH THE AGES

*by* ARTHUR D. HOWARD, Ph.D.

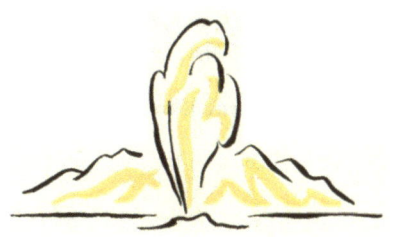

**COLUMBIA UNIVERSITY PRESS**

NEW YORK : 1938

COPYRIGHT 1938
COLUMBIA UNIVERSITY PRESS, NEW YORK
FIRST PRINTING, JUNE 1938
SECOND PRINTING, AUGUST 1938

*Foreign agents:* OXFORD UNIVERSITY PRESS, *Humphrey Milford, Amen House, London, E.C. 4, England,* AND *B. I. Building, Nicol Road, Bombay, India;* KWANG HSUEH PUBLISHING HOUSE, *140 Peking Road, Shanghai, China;* MARUZEN COMPANY, LTD., *6 Nihonbashi, Tori-Nichome, Tokyo, Japan*

MANUFACTURED IN THE UNITED STATES OF AMERICA

*To* DR. C. MAX BAUER

*Chief Naturalist of Yellowstone National Park*

*The author is indebted to the following for permission to reproduce the photographs: to the U. S. Army Air Corps for* BEARTOOTH MOUNTAINS, MONTANA; *to Professor Douglas Johnson of Columbia University for* YELLOWSTONE PLATEAU; *to the U. S. Geological Survey for* TABLE MOUNTAIN, ABSAROKA RANGE; *and to Haynes, Inc., for* RIVERSIDE GEYSER, *for* MORNING GLORY POOL, *and for* MAMMOTH HOT SPRINGS TERRACE. *Acknowledgment is made as well to Dr. Edward Hodnett of Columbia University Press for the pen-and-ink drawings. The author desires also to acknowledge his debt to Professor Richard M. Field of Princeton University and to Professor Douglas Johnson for aiding and encouraging him in his researches in the Yellowstone region.*

# FOREWORD

### By C. MAX BAUER, Ph.D., Park Naturalist

OUR national parks are not only scenic wonderlands, but many of them contain unique examples of the work of nature and are therefore of great interest to the student and to the scientist. For this reason, the National Park Service encourages competent investigators to study the natural features in our national parks and through their research work to contribute to the knowledge of the origin and history of these interesting phenomena.

It is the function of the Naturalist Division of the National Park Service to coördinate and correlate the scientific material and facts gathered by the many scientific workers and to make the knowledge available and understandable to the public.

For the past several years, many investigators have been interested in Yellowstone National Park in order to obtain a better understanding of the events of its past history which have brought about its beautiful canyons, waterfalls, lakes, hot springs, geysers, and accumulations of volcanic materials. Dr. Arthur D. Howard has spent five seasons studying the geology of the Grand Canyon of the Yellowstone. In the following pages he has told in popular language the story of Yellowstone through the ages, and I am sure that the reader will find it fascinating, enjoyable, and profitable.

By courtesy of the U. S. Army Air Corps

BEARTOOTH MOUNTAINS, MONTANA

GALLATIN RANGE

# I. TODAY

A BREATH-TAKING ride up the precipitous, zigzag road brings us to the top of Mt. Washburn. Here on the bare mountain crest we breathe more freely. The air is thin and clear; there is a temporary lull in the wind. Half a mile below us lies the Yellowstone Plateau, its surface 8,000 feet above the sea. And as though on a great relief map, mountains and canyons and lakes and rivers are spread on all sides, covering an area almost as large as the state of Connecticut. From our lofty vantage point we gaze across primeval wilderness penetrated by only a few roads and known intimately to but a handful of men.

Down among those trees live elk and antelope, black bears and grizzly bears, buffalo and moose, and the rocky slopes below us are favorite haunts of bighorn sheep. The animals are there, as they have been for centuries. Blackfoot, Crow, and Shoshone hunted and trapped them, ages before white men dreamed of exploring the wilderness. Generations of Indians, long separated from their kinsmen in Asia, caught fish in the streams of Montana and Wyoming. Many of them, in their long treks across Yellowstone, camped along these rivers spread

NOTE: See the end of the book for geological diagram and map.

in a network below us; they knew the evil spirits of geysers and scalding springs, and their steps were hastened by fear. Only the lowly, timid Sheepeaters made Yellowstone their permanent home; it offered refuge from the warlike tribes beyond the mountain passes.

The first white man to watch a sunset over Solfatara Plateau was John Colter. He was a member of the Lewis and Clark expedition. With forty-five others, he paddled up the Missouri and Jefferson rivers, crossed the Rocky Mountains, and endured the winter of 1805 and 1806 in Oregon. On the return journey, while the expedition was encamped at Fort Mandan in North Dakota, Colter asked permission to return to the wilderness. Trappers had told him stories of game and fish in a land where white men seldom hunted. Perhaps they told him of savage Indians, of peril and adventure. This land now belonged to the United States; we had bought it from Napoleon Bonaparte in 1803 at the time of the Louisiana Purchase Treaty, and Colter wanted to see what we had bought. So into Wyoming he went. For a long time the boundaries of Yellowstone held his interest; yet he was not eager to plunge into the wilderness beyond. But in an Indian fight south of Yellowstone, on the west side of the Teton Mountains, Colter was wounded in the leg. The shortest route to the fort at the mouth of the Bighorn River led through the heart of Yellowstone, and alone and wounded, he struggled across the wildest country he had ever seen. He found there all that he expected—fish and game, pine forests and meadowlands—and all that he had known before. But he had never seen springs delicately tinted and steaming hot, a canyon drenched in yellow and gold, water that shot up from rocks in shimmering plumes a hundred feet and higher into the air. Neither had any other white man seen these things; and when Colter several years later wandered into St. Louis, he and his tales became the butt of ridicule. Tall stories flourished on the frontier. Fish and game and Indians, they were commonplace; geysers and hot springs were incredible. And the first name men gave to Yellowstone was "Colter's Hell."

More than a century has passed since then. We stand above the trail which Colter must have followed, along the Yellowstone River and over the flank of the Washburn Range to Tower Falls. Those are the very trees through which he wandered— a sea of evergreens dominated by slender lodgepole pines and interspersed with stands of whitebark pine and spruce, fir and limber pine. We can smell their sharp fragrance. In spring and summer, the little stands of aspen dot the map with lighter green, but aspen leaves turn yellow in the fall, and then the forests are flecked with gold. Meadowlands break the line of forests; elk and antelope are browsing there, just as they were when John Colter hunted them.

Now we can see that the plateau is not everywhere flat. Far to the southwest, spouts of steam mark the site of one of the main geyser basins, but there are other geyser basins which we cannot see from here. Hidden, too, are the troughs within which nestle Heart Lake and Shoshone, Lewis Lake and Beach. There in the south is the broad lowland of Hayden Valley, where the Yellowstone River, dark with reflections of evergreens, meanders peacefully along its eastern margin. And still farther south, the blue waters of Yellowstone Lake fill a vast hollow in the southern plain. But if there are moats in the plateau, there are towers too. Flat Mountain and Purple Mountain, Stonetop and Gibbon, all are lower than our summit, but they stand hundreds of feet above the surrounding plateau.

Many large rivers have their source in and around Yellowstone. Far to the south, a tiny stream flows down the west side of the Continental Divide, wends its way westward gathering tributaries as it goes, and finally empties into the Columbia River and the Pacific Ocean. Its serpentine wanderings long ago earned it the name Snake River. Madison and Gallatin rivers in the west, the Wind River in the southeast, and the Shoshone in the east, flow into the Missouri and the Atlantic.

YELLOWSTONE PLATEAU

Largest of the rivers of Yellowstone is the Yellowstone River itself. It wanders into the Park in the southeast, and we watch it flow gently northward, silently mingle with the placid blue waters of Yellowstone Lake, and then meander across Hayden Valley. But once beyond these tolerant banks, the Yellowstone turns savage. It swirls northeast and plunges over a cataract, drops more than a hundred feet in storming silver froth; over and down again, this time three hundred feet. Savage still, it rushes through a narrow gorge—the Grand Canyon of the Yellowstone, "cameo of canyons." The canyon walls are carved into a filigree of ravines and pinnacles, gullies and spires. Yellow and white, red and green, orange and sunset pink, the walls slope upward from the raging torrent. Relentless erosion and highly heated waters and gases have wrought incredible splendor in the canyon. Erosion has given it its present depth and etched its walls, and the hot waters and vapors have brought out the brilliant coloring of the rocks. But heedless of the beauty of the canyon, the river plunges onward seeking escape from the confining walls. At Tower Junction, joined by its greatest tributary, the Lamar, it leaves the canyon. Tranquil again, it flows through less spectacular gorges, and at Gardiner, Montana, north of Mammoth Hot Springs, it glides from the Park.

We have been looking down. Look up now—for though we are 10,000 feet above the sea, we can still look up—and see in the distance the walls of Yellowstone, magnificent as the castled sides of the Grand Canyon. In the northwest, beyond this mountain where we stand, and beyond the other bare-crested knobs of the Washburn Range, barren slopes rise sheer against the sky. That line of pyramidal peaks is the Gallatin Range. Majestic we see it, a long range, rocky brown merging into purple among the summits. Now the sun encircles the mountains with a faint halo, and the lower slopes take on darker shades of green. Only twenty miles of the Gallatins belong to the Park, but the mountains stretch far northward into the state of Montana.

Now turn right, and face the northern boundary of Yellowstone. Grey and somber looms the Beartooth Range, untouched by the sun and dark as the bank of clouds that hangs above it. For forty miles we see nothing else—only the serrated peaks that fade into evening mists, canyons that look from here like jagged clefts in the grey rock.

The sun is sinking; soon those western mountains will hide it. While we have time, look east, where the Absarokas shield Yellowstone from the outside world. The Absarokas, layer on layer of flat, dark rock, rise to flat and rocky summits. They stretch almost fifty miles to the east, forbidding as the Beartooth Range in their somber, brown-purple splendor. No wonder that attempts to enter Yellowstone from the east were so long thwarted.

But there is much to gaze at, and soon it will be dark. Look south, into that region of high, rolling hills. Those are the northernmost foothills of the Wind River Range; they form the mountain border in the south. Dense forests mantle the rounded summits and clothe the smooth slopes. No one knows much about this hilly wilderness; only a few hunters and trappers penetrate that wild and inaccessible country.

And west of the Wind River Range, across the fertile valley of Snake River, stand the Tetons, the grandest mountains of them all. Their northern slopes outline the southwestern rim of Yellowstone. The peaks are cathedral turrets that reach 7,000 feet above the lowlands—the loftly spire of the Grand Teton soars almost 14,000 feet above the sea.

We have followed the outline of Yellowstone in an almost complete circle, but as the Tetons slope to foothills and meadows, we see a gap in the wall. Yellowstone Plateau squeezes through the gap, and ends, a few miles outside the Park, on a forested precipice that overlooks the lower plains of Snake River.

Yellowstone is wilderness surrounded by wildness. It is a mountain-encircled plain 8,000 feet above sea level, a land of strange phenomena and a haven for animals. As we stare about at the tall mountains on all sides, still white with the winter's snow, there come to mind the words written in tribute to the loveliness of the Vale of Kashmir—"an emerald in a setting of pearls"—and we find those words peculiarly appropriate here. Now that the sun has set behind the Gallatins, we may descend from our mountaintop and turn our thoughts to the origin of this unusual region.

## II. BIRTH OF YELLOWSTONE

Turn whichever way you will in Yellowstone, somewhere near at hand, in the walls of the Grand Canyon, in the steep face of Sheepeater Cliffs, in the sheer slopes of the Gallatin Range, you will find jutting ledges of rock. Each layer of rock, whether it be an old lava or an ancient limestone, has a story to tell: the lava tells of an ancestral volcano or of a yawning fissure vomiting volcanic rock; the limestone, of a long-forgotten sea. Read the story of each rock, compare it with that of its neighbors, arrange them in order, and the complete story of Yellowstone emerges.

Half a billion years ago, the greater part of North America was a barren and dreary plain. There were no trees, there was no grass; no life, because there was no food. Along the Atlantic coast towered a mountain range called Appalachia, which stretched from Florida to Newfoundland. Its eastern slopes disappeared under the waters of the Atlantic far beyond our present coast line, and its western slopes stretched to the foothills of our modern Appalachians. The roots of these ancient mountains are exposed on the surface today over a great area, and are known to underlie the coastal plain. But where our Appalachians now stand, a shallow trough once paralleled the slopes of Appalachia. For eons of time it lay there, a long and narrow basin, and during eons of time streams carried into it mountain sediments. Grain by grain, Appalachia was brought down into the trough, and the trough was crowded with sediment. Then, centuries later, the crust upheaved, and the great thickness of sediment which had been carried down was crumpled and thrown into folds to form our Appalachians—humbler than ancient Appalachia, but standing high above the severed roots of that old mountain mass.

The same thing happened on the western coast. There the range Cascadia dominated the Pacific. Along its eastern slopes ran a narrow trough, and in the trough were spread sediments from the peaks of Cascadia. Most of the sediment was piled at the foot of the mountains, where the swift streams were first checked; but the smaller particles were carried farther and farther. And so we know where the highlands once stood, for although the sediments have long since turned to rock, coarsening texture to the westward tells us that the mountains which were their source lay in that direction. As Appalachia vanished, so did Cascadia, for at the same time that the eastern sediments were being upheaved and crumpled into new mountains, so were the western. Mountains grew where troughs had lain, and of Appalachia and Cascadia only the roots remained.

Half a billion years ago is almost too remote to imagine. Yet the cycle of uplift and downfall continues, and will continue for incalculable years beyond us. Mountains are being washed by streams today. Spring thaws swell the streams, and summer drought parches them. So it was when Cascadia looked down upon the Pacific coast. Sometimes the streams in their vigor carried gravel far beyond the mountain front; at other times the gravel was laid down at the foothills, and the weakened streams spread only sand and clay over the basin floor. And so, layer on layer, the sediments at the foot of the ancient mountains grew thicker—a bed of gravel, a sheet of sand, a blanket of clay, and then, perhaps, more gravel. And although these sediments were originally spread out in nearly horizontal sheets, today they rise at steep angles and bear witness to the struggles of the past: relentless destruction of the old, inevitable birth of the new.

## Earliest Years

When Cascadia and Appalachia were in their prime, Yellowstone was a low, featureless plain, and not so luxuriant a plain as this Central Plateau on which we stand. There were no geysers spouting then, no beautiful lakes and waterfalls, no deep canyons and lofty mountains.

In these earliest years, the north end of the Cascadian trough lay submerged under the waters of the Arctic Ocean, and the south end was swept by the Pacific. Ancient sea creatures lived and died in these waters, and their remains were entombed in the sediments washed into the seas from the slopes of Cascadia. Remains of former life are known as fossils, and fossils of sea creatures tell us that regions now dry once lay under water. So it is with the Alps, where fossil sea shells are found in the rocks of the highest peaks. To simplify the telling of our story, we shall refer to sediments which are laid down in the sea and so contain fossils of sea creatures as marine and to those that are laid down on dry land as continental.

Sometimes the ocean waters covered only small areas at the ends of the trough, but at other times the seas swept widely from north to south. We know this by the spread of the marine sediments. The sediments are in layers; if we studied one layer, we could say: Three hundred million years ago the ocean swept inland for a hundred miles. But if we examined the next layer, we might know that a few million years later the oceans inundated the entire trough from one end to the other. From these sedimentary layers we have learned the story of Yellowstone's youth, for Yellowstone was part of the vast western region submerged by these vacillating seas.

At the time the first sediments were swept down from the mountains, the waters of the Arctic and Pacific oceans were advancing into the trough. Perhaps the oceans were rising, perhaps the continent was sinking. No one knows. But age after age the invading seas drew closer. At last they merged, to form a narrow seaway between the foothills of Cascadia and the interior plains of North America. No fish swam in the waters of that ancient sea, but there were shelled animals similar to those of our modern seas and strange little segmented creatures which have long since perished from the earth.

Thousands of years later, the seas withdrew. They laid bare the floor of the trough. And over the floor the mountain streams spread other layers of sediment.

Even after that, the oceans had wanderlust. For hundreds of millions of years they played their game—advance and retreat, flood and drain. Perhaps their second journey eclipsed the shore line of the first; perhaps the third fell short of it. Sometimes the combined waters of the Pacific and the Arctic swept across the continent to meet Atlantic waters in the eastern trough of Appalachia. At such times the remains of sea creatures were spread from one side of the continent to the other. While the seas stayed in bounds, continental sediments, barren of fossils, were spread over the lands.

For countless ages Cascadia was preyed upon by the streams which it nourished, and eventually the trough at its base was filled with sediment. Upheavals of the crust have long since lifted these sediments so high into the air that today we can see them in the mountain ranges of the west, from Alaska to Mexico. Here in Yellowstone they lie exposed to the north of us in the towers of the Gallatins.

If thousands of feet of sediment came down from ancient Cascadia, how high must those mountains have stood—higher than the Rockies, high as the Swiss Alps. How many million years will pass before the Rockies and the Alps are plains and our descendants compare new mountains with those of the twentieth century?

## Earliest Life

Till now, Yellowstone's history has been told in monotone, an unexciting tale of endless invasions by a restless sea. Once dry and barren, then flooded by the sea; arid again, then washed by another flood. For hundreds of millions of years, the wandering oceans had their way. Yet this was a time of miracles, for the land was bearing fruit; the wastelands were clothing themselves in vegetation. And out of the sea, to the promise of food and luxury, crawled the ancestors of all the animals that later roamed the land and flew the skies. Slimy creatures at home on land or sea came first; later, weird finback brutes roved farther inland. And through the air droned giant insects, many times the size of their modern descendants. King of them all was a dragonfly with a wing spread of twenty-nine inches. These were the creatures that knew Yellowstone in her infancy.

## Upheaval of the Crust

After this long period of quiet, North America rebelled. Parts of the crust were twisted and shattered; volcanoes spat lava and broken rock; and the flat-lying sediments in the troughs of Cascadia and Appalachia were crumpled and flung upwards into mountains. The long eastern trough was now a gigantic mountain range, but in the west, though some mountains were born, other areas still lay flat. To balance the new highlands, the beveled roots of Cascadia and Appalachia settled, and there were coastal plains instead of coastal mountains.

The shift of highland to lowland bewildered and destroyed some of the creatures who had lived in the quieter days of the earth. Luxuriant forests gave way to barren wastes. Mountain glaciers covered soil that had once nourished lowland meadows. Winds and ocean currents were changed by the continent's rise; flowers and trees were starved, and deserts spread sand over the parched roots. Only a few hardy families of animals survived; the finback reptiles, and others whose lives had depended on warmth and abundant vegetation, vanished.

The story of the world's youth might have remained a secret, except for that crumpling of rocks. Every horizontal layer of sediment conceals the layer below. Without the upheaval and folding of layers, we should never have known that the seas swept in or how far they covered the land; and we should still wonder what plants and animals lived on the ancient earth. But erosion has bared the edges of the crumpled rock layers, and we can read the earliest pages of history.

DINOSAURS & FIRST BIRD

## III. YOUTH OF YELLOWSTONE

DAYBREAK on the road between Mammoth Hot Springs and Tower Falls. The mountains rise east, west, north, south. Their slopes and crags and summits seem as enduring as time itself. Yet ever since the world was young, mountains have risen only to be worn down again to their roots. So it was with Cascadia and Appalachia; and so it was with the ancestral Appalachians and the ancestral Rockies, newly risen from the sediments of the earlier troughs. During the span of the following hundred million years, the Appalachians, towering to Alpine heights, were filed to sea level and again upheaved. And the scattered ranges of the Rocky Mountains suffered similar fates. Along the coasts, the encroaching seas submerged the roots of Cascadia and Appalachia, but the interior of the continent remained a vast, flat plain, and neither the Great Lakes nor Hudson Bay relieved the monotony of that endless expanse.

### RED DESERT

Yellowstone's youth was without promise. The region lay parched under a burning sun; brilliant red sediments simmered over its bleak and treeless wastes, and the air above danced with heat waves. It is the same red sediment that patches hills to the east of Mt. Everts, which to the north of us lifts its flat summit. And the whole continent simmered thus;

red salt-laden sediments were laid down from coast to coast. The red valley of the Black Hills of South Dakota and the red meadows of New Jersey are tinted by these ancient sediments.

## Sundance Sea

Yellowstone today abounds in sparkling rivers and gemlike lakes. The lakes stir restlessly this morning. The roar of waterfalls defies the wind. Yellowstone River, sullen under the lowering skies, flows northwest of us, meeting its tributaries—Cottonwood Creek, Hellroaring Creek, and the Little Buffalo. Yellowstone is humid compared to the sun-blistered wasteland of a former day; and yet, for all its lakes and rivers, it is arid compared to what it was when Arctic waters swept from the north again and drowned the red wastelands of Yellowstone under a flood hundreds of feet deep. Because the clues which led to the discovery of the ancient flood were first recognized near Sundance, Wyoming, men refer to this long arm of the northern ocean as the Sundance Sea.

Strange animals swam the waters of this Arctic embayment. A monster twenty-five feet long, with a turtle body and the neck and tail of a serpent, swelled its waters. Woe to the unwary fish that came within reach, for clumsy though he was, the monster struck with the speed of a cobra. Tribes of sea creatures fought and frolicked here, and gave birth through the years to monsters with longer tails and huger bodies. For thousands of years they owned Yellowstone, which lay at the bottom of the Sundance Sea.

## The Dinosaurs

There were other reptiles, beyond the shore lines of the Sundance Sea. They were the dinosaurs. As the sea slowly drained back to its ancient boundaries, the dinosaurs advanced. They took possession of the old sea floor, finding the level, marshy plain to their liking. Some of the beasts, three times as tall as a tall man and twice as long as they were tall, walked upright on their powerful hind limbs. Yet these two-footed carnivores that reared twenty feet above the ground and measured forty-five feet from snout to tail were small. There were other dinosaurs, four-footed, vegetarian, with ponderous, elephantlike bodies and long, slender necks and tails, and they reached eighty feet in length. Huge though these creatures were, they had only their crushing weight to defend themselves with. They hid, most of the time, in the murky swamps where the fierce flesh-eaters dared not come. But ever so often, one of these gigantic beasts would wander out on dry land, or one of the more daring flesh-eaters would invade the swamps; then the plains and the marshes ran red with blood. There was another weird beast, with armor of bony plates on his back and with a powerful spike-studded tail, but this creature was an inoffensive vegetarian, and like the little porcupine of today, used his formidable armament only in defense. Batlike monsters lived in the air, and in later generations spread their wings for twenty-five feet across the sun. The dinosaurs lived not only in the place we call Yellowstone, but all over the earth. The first remains were found in England, but there were dinosaurs in New Jersey and Mongolia and the South of France, on every continent, from the Gobi Desert to the interior of Africa. It was a world of dinosaurs.

## The Last Flood

The dinosaurs' day was a long one, but it was interrupted. Once again the sea swept over the land, flooding the continent from the Arctic Ocean to the Gulf of Mexico. The swollen waters stretched from the western mountains to where the Mississippi River now flows, a distance of a thousand miles. The flood demolished the dinosaurs' paradise and brought back its own citizens, whose ancestors had played in the Sundance Sea. The new generation was well fed; within the aquarium that was Yellowstone lived millions of little shelled animals. The sea reptiles gobbled them up and grew fat and contented.

Then once again the sea ebbed from the land. Perhaps the trough was becoming filled with sediment from rising mountains to the west; perhaps the entire continent was emerging from the sea. Once more the marshland bred dinosaurs. The ruined kingdom was taken over by descendants of its former rulers. And if some of the old monsters had vanished, there were new types to take their places. One of them was a rhinoceroslike creature with a fan-shaped collar at the back of his neck. He was larger than the rhinoceros, and his collar was a bony frill that shielded his body.

And the plant world became more beautiful as flowers sprinkled the landscape with color and the eternal evergreens gave way to trees whose green leaves each autumn turned red and yellow. Flowers, seeds, and luscious leaves grew in profusion, and furnished food so rich and wholesome that a race of tiny animals, hitherto inconspicuous, began to increase rapidly in size and number. These were the mammals, warm-blooded, milk-bearing creatures, who found dinosaur eggs to their liking, and by their depredations may have threatened the existence of their huge neighbors.

## Molding of the Landscape

For almost half a billion years, Yellowstone had been buffeted. The ocean had flooded and drowned her. She had been exposed as a red and dreary wasteland. She rebelled.

HORNED DINOSAUR

Amid the thunder of distant volcanoes, the earth heaved and writhed and shook its crust of sediment. The crust folded and broke, and Yellowstone emerged. She flung up piles of sediment and formed the Gallatin Range—mountains which climbed toward the sky much as they do now. Then Yellowstone barricaded herself on the north. She crumpled more sediment into folds and shaped the mighty Beartooth Range. She looked into the future and formed her eastern barrier—the ancestors of our Absaroka Mountains. And in the south she heaved up the snow-clad summits of the Wind River Range.

The southwest border of Yellowstone remained for a while a low and broken basin rim, where now, miles away, we glimpse the Tetons, topped by the craggy head of the Grand Teton. And the Washburn Range waited for birth, and centuries passed before the Red Mountains grew up between two placid lakes. Where we see the broad, level shelf of Yellowstone Plateau, there was then only a deep basin, its floor littered with sediment.

### End of the Reptile Dynasty

The dinosaurs were doomed. Cold-blooded reptiles must live in tropical climates, and when the climate of the lands cooled in response to the upheaval of the crust, the dinosaurs declined. They had grown, through the centuries, to enormous size, but their brains remained tiny and undeveloped. They could not meet change. And Yellowstone favored the mammals. Infinitesimal beside the great reptiles, they flourished among the new plants and trees. The giants of Yellowstone disappeared—the dinosaurs, and the dragons whose wings obscured the sun, and the monsters of the Sundance Sea. Yellowstone's upflung mountains were hills to them, but the mountains still stand, and the dinosaurs are gone.

ANIMALS OF THE VOLCANIC PERIOD

## IV. AGE OF VOLCANISM

WE SAW the Absarokas from Mt. Washburn in early morning, and now from Central Plateau we stare up at the distant peaks glittering in the noonday sun. We have tried to imagine Yellowstone without the Absarokas, and we have thought back to the earliest ancestors of this eastern boundary. Before these present mountains grew, there was still another period when Yellowstone had no rugged eastern boundary, for erosion wore down the old mountains almost before they had reached their full height. Streams gnawed at the slopes and carried the debris to the rugged floor of the Yellowstone Basin. The basin was filled with waste from the ancestral Absarokas and the other encircling mountains, and the surface rose thousands of feet above the old lowland. Then the climate grew more moist. Streams were replenished, and struggled with new vigor to cut through the sedimentary fill. It was not such a long time, if we think of a century as our smallest unit, before the basin was cleared again. But in the south, where we can see only the summits of Chicken Ridge and the bearded foothills of the Wind River Range, remnants of gravel remain on the mountain slopes—the only memorandum Yellowstone made of that historical paragraph.

TABLE MOUNTAIN, ABSAROKA RANGE

By courtesy of the U. S. Geological Survey

## Growth of the Absarokas

When the earlier range was worn down to a low mountain line, the modern Absarokas took up their climb to the clouds. Here and there thunderous explosions shattered the crust of earth. Each explosion built up around its vent a gigantic pile of debris—rock fragments many feet across, mixed with fine volcanic ash. Every violent explosion spewed out a layer of broken rock, and every quiet eruption poured liquid lava over the layer. Volcanoes grew, exploded, and grew again. Only our loftiest volcanoes, Vesuvius in Italy, Mauna Loa in Hawaii, Cotopaxi in Ecuador, can equal those ancient mountains.

But even the fury of volcanoes was helpless against the eroding streams. The centers of eruption shifted, and the volcanoes standing above the dead vents were carried away piecemeal. New deposits buried the old; new volcanoes rose; year after year accumulated the layers of broken rock and lava that built the Absarokas.

We can follow every step in the life of the eastern mountains. The largest rocks fell nearest the vent that catapulted them forth; the finest ash flew farther. Even though the cone of a volcano has vanished, we can trace back, from delicate ash to largest rock fragments, and there find the ancient point of explosion. Sometimes a monolith—a single shaft—of rock hundreds of feet high rises to remind us of the old vent. Lava which solidified in the volcanic tube fought mightily against erosion. It pitted a passive strength against the nervous war of streams, and resisted the force which swept away the volcanoes. Even though these monoliths are eventually destroyed by erosion, their roots bear testimony to the volcanoes of the past. Sometimes, too, volcanoes ruptured themselves in fury, and broke into a series of fractures that radiated out from the center. Then lava bubbled up from the earth to fill the fractures with vertical sheets of rock. These stood silent too under the fretting of streams; and though the volcanoes are gone, we find the narrow sheets of rock still standing, stretching outward like the spokes of some huge wheel.

So the Absarokas erupted and grew. And their growth left Yellowstone tenantless. The basin was bombarded with volleys of shattered rock, and floods of boiling lava spread wide. Dust-choked, sunless, crushed under rock and drowned in burning lava, Yellowstone was left to herself. Few creatures could live there while the Absarokas grew.

Yet even the volcanoes tired at times. Then they lay for centuries dormant, husbanding their strength. And while they slept, erosion labored. Rocks were rolled away and lost their jagged edges in the streams. Such deposits of rounded pebbles and cobbles we know as gravel, and when time, in its patience, has welded the gravel into a resistant rock, we call it conglomerate.

The greatest memorials of the periods of volcanic quiescence are the petrified forests. While volcanoes rested, new soils grew over the debris in the basin. Trees rooted themselves in the earth and grew gigantic, ten feet and more in diameter. The volcanoes waked and buried the forests; volcanoes slept, and new centuries found new soil and new forests. On the north flank of Specimen Ridge—north of us, and out of sight—about a dozen of the primeval forests are buried, one above the other. And after the trees were buried, underground waters, percolating through the soil, dissolved their woody tissue and put in its place silica, harder than wood, which has withstood erosion. Thus Yellowstone preserved her ancient forests.

We are surrounded by the deposits of the volcanic age. When we drove through the eastern entrance of the Park, we were driving through the heart of the Absarokas. Remember the layers of angular rock that stood out in the road cuts and on the faces of cliffs? Those volcanic fragments are called breccia. They were flung there while volcanoes were awake and wild. But while volcanoes slept, many angular fragments were rolled about and rounded, and formed gravel, which rode with the current of streams to the lower slopes, and eventually formed layers of conglomerate. The fine-grained rock that we saw elsewhere on Absarokan flanks is volcanic ash, once blown from the open mouth of a crater. And at Sylvan Pass, where we thrilled to be eight and a half thousand feet above the sea, we saw other lavas that flooded the region millions of years ago.

It is not so hard for us to imagine those days of fire and crater smoke when we can still see their handiwork. And a span of a million years is not so long, geologically speaking. It took tens of millions of years to build the Absarokas, but the massive pile that was built is more than 10,000 feet thick. Part of it, carved into hills and valleys, is buried under our feet in the depths of Central Plateau.

### BIRTH OF THE TETONS AND THE WASHBURN RANGE

Volcanoes exploded to build the Absarokas. At the birth of the Tetons, during the volcanic era, earthquakes rocked the region at close-spaced intervals and helped frighten away the few intrepid creatures who had wandered into the inferno from the safer world beyond the mountain rim. But the volcanoes and earthquakes worked to a purpose. The earthquakes were only growing pains of the Teton Range. The crust of the earth in the southwest was shattered and uplifted; and amid the violence of earthquakes, the Grand Teton flung up its im-

TETON MTS. JACKSON LAKE

perial, snow-shrouded head. Incredible birth—a trap door of the earth rose slowly, a few feet, a few tens of feet every hundred years. Upward jerks that shook the earth, a pile of massive rock that grew with the majestic, imperturbable tide of evolution. It grew until it stood 7,000 feet above the lowlands. Its smooth face was scarred by erosion and separated into a series of peaks. The heaving of the earth lifted rocks of the valley floor high into the air—they are there still, far up on the western flanks of the range. And some of the rocks are marine. Centuries ago, they lay undersea; now they stand almost on the crest of one of the loftiest mountaintops in Wyoming, this state of lofty mountains. For though it is hard to believe, these Tetons that pierce the sky were once flush with the floor of the Jackson Lake lowland.

Earthquakes roused the Washburn Range at about the same time that the Tetons were rearing their shaggy heads. You remember, as we stood on Mt. Washburn and stared at the sunset beyond the Gallatins, we saw that the Washburn Mountains form a mammoth horseshoe, open to the northeast. And we saw that there were three distinct sections of the horseshoe, two long ranges which run northeast to southwest and the third running north and south to connect their western extremities. We know now that the Washburns are smaller imitations of the cloud-bonneted Tetons. Slowly, Teton-like, the three segments of the horseshoe rose above the debris-choked lowlands of Yellowstone. The north and south segments rose with the open trap door facing south, but the western section left its trap door open to the east. Although not so lofty as the Tetons, the Washburn Mountains were giants in those days, for foothills increased their height; now the foothills are buried under layers of lava. The surface from which the mountains rose is buried under these same lavas. But near Mt. Washburn, the Yellowstone River has cut its canyon deep into the lavas, and the old topography shows. The surface there was low and rolling, millions of years ago, and the streams that drained it merged to flow northward into a broad valley. Now that valley is a trysting place for the Lamar and Yellowstone rivers, but the ancient stream escaped to the north through the enclosing mountains along almost the same path that our present Yellowstone follows. Geologists call it the Ancestral Yellowstone River.

## Early Lava Flows

The Ancestral Yellowstone River flowed undisturbed through its valley for centuries. Then lava crept up through fractures in the earth and flooded across the plain. The floor of the river valley was buried first under black rock, later under a lighter-colored flood. But along the main highway at the foot of Crescent Hill in the north, where we can see those early flows today, the black rock has weathered brown, and the lighter rock has been encrusted with layers of almost black plant growths. Only on surfaces just broken can we see the original coloring. Once the black-brown and brown-black rocks lay at the bottom of Yellowstone Valley, but the river has since cut hundreds of feet lower, leaving remnants of these old lavas perched on the valley sides. These lavas meant little to Yellowstone, for they covered only small areas in the north.

And then more lava—black lava, which we call basalt—rose from some subterranean reservoir to the surface and spread over the lowlands. We have seen that lava. From the dirt road around Bunsen Peak, south of Mammoth Hot Springs, we saw in the valley of Gardiner Canyon the thin flows of basalt which gushed over the ground, millions of years before geologists lived to call it basalt. And we remember Overhanging Cliff, the beetling black-rock cliff under which crept our road to Tower Falls, and the gigantic columns across the canyon, like palisades surrounding a besieged city—these too were formed in the earliest days of flowing lava. Lava flowed into a valley, whose floor was covered with sand and gravel. In that valley played creatures who knew Yellowstone only as a friendly, quiet place. The bones of an ancestor of a horse have been found there, undisturbed for millions of years, buried in gravel that once cradled a stream. Those animals lived in the interludes of history, when volcanoes and earthquakes merely whispered. For years the prehistoric beasts grazed in the meadows and stalked among the pines. Fat and well-fed on the lush vegetation, they were safe from their carnivorous neighbors. How could they tell that next year, next day, volcanoes would bury them in floods of lava? How could they know that where they grazed, a river would carve a gorge, that men would call it Yellowstone, and that the bones of some who had roamed this place would lie hidden under rock and debris, below the jutting canopy of a cliff, until geologists millions of years later came to dig them up?

The first lava flows had finished their work, and ceased. Once more the patient streams began to clear the lowlands, digging and gnawing at the frozen flows of lava. They cut valleys through the lava and down into the rock beneath. They sliced below the base of Overhanging Cliff and gouged a chasm 400 feet deep. And in the north, the valley of the Ancestral Yellowstone, its floor buried under hundreds of feet of the black basalt, was swept nearly bare of the encumbering lava. Only a few remnants of the early lava flows remain in the valley— the citadel of Overhanging Cliff, the black cap of Junction Butte, a few black patches scattered against the grey valley slopes.

## THE GREAT LAVA FLOOD

High noon is sunny and placid here on the plateau. But things were not always so. Once white dust powdered the sky, clouding the sun and drifting down over the landscape like snow. Volcanoes roared, the earth yawned and cracked, and Yellowstone became a molten sea of white-hot lava, fifty miles across, 2,000 feet deep. The shore lines of this awesome flood seared the feet of the encircling mountains. The lava invaded the mouths of valleys and changed them into bays; and the rocky spurs that separated the bays were like gnarled fingers thrust into the sea. The streams coursing down the mountain valleys met their fate in this simmering ocean—when water touched those burning shores, steam rose and water vanished. Lava breathed more steam into the air, and the steam rose in a dense, impenetrable fog which billowed over the burning sea and would not withdraw. It rose upward into higher, colder atmosphere and was condensed into rain. Yet no rain fell on Yellowstone. Long before the waters could touch earth, the heat of the molten lava forced them into steam, and up they rose, wraithlike, on their never-ending journey. Like lost souls the waters of Yellowstone hovered between heaven and earth.

Rocked by volcanoes, shrouded in scalding steam, saturated with poison gases, buried under lava, Yellowstone harbored no living thing.

But the centuries passed. Lava cooled to rock. Today most of the Yellowstone Plateau is formed of that once-molten lava. We have seen the rock in most of the cliffs and most of the road cuts in the Park. Geologists call it rhyolite.

In some places the lava cooled rapidly to form the shiny black rock called obsidian. We saw it in the freshly broken surfaces of Obsidian Cliff, whose grey, weather-beaten face rises from the road between Mammoth Hot Springs and Norris Geyser Basin. Red men made countless pilgrimages into Yellowstone in quest of this hard, shiny rock so easily flaked into arrowheads and spear points. The foam and froth of the lava sea congealed most rapidly and formed a porous rock, some of which mantles the summit of Obsidian Cliff. This is pumice, a rock highly prized in modern times as an abrasive.

When the lava cooled, the waters of Yellowstone descended from their long suspension in the air. But the plastic surgery of volcanic forces had transformed the region. There was no longer an irregular basin. A level plain stretched westward beyond Yellowstone's horizon. Above the frozen sea of lava, the higher parts of old Yellowstone rose like islands—the Washburn Range highest of all. In the south, Red Mountains rose, new and lovely, tinted a faint red in the sun. Mt. Sheridan, reaching up above the other peaks, may be one of the volcanoes which smothered Yellowstone.

Before the shore line of the rhyolite sea had reached the northern part of the Park, other flows of basalt were drowning the valley of the Ancestral Yellowstone River. At the mouth of Blacktail Deer Creek, the flows poured out so swiftly that now they lie stacked, one above the other, like pancakes. But elsewhere in the canyon, north of Tower Creek where we once stood, the flow of lava was irregular. It flowed and spread and waited for years to flow again. While it waited, streams eroded it and covered it with debris brought down from upstream; and perhaps centuries later, other lava covered that debris. On the east wall of the canyon, 150 feet of gravel rests on an early lava flow, and the gravel is covered by a later flow. Do you remember this later flow with its stockade of columns? The Palisades, we call it. The explanation of its appearance is simple:

When water freezes, it expands. We put alcohol in our car radiators to prevent the water from turning to ice and bursting the radiators. But molten rock contracts when it freezes, and cracks like a dried-up mud flat. If the lava cools rapidly, then the cracks are close together and the columns of rock left in between them are thin. If the lava cools slowly, cracks are wide apart, and the columns of rock are thick like those in the lower part of Overhanging Cliff. If streams had worn the cliff down to the layer of thick columns, then a uniform wall of pillars would remain, like those across the canyon.

All during the time that basalt was flooding the northern lowlands, the rhyolite sea was rising higher and spreading farther to the north. It spread to Tower Falls. And there the flood squeezed through the gap between Specimen Ridge and the northern spur of the Washburn Range, and, swelling, poured into the Ancestral Yellowstone Valley. The valley was filled from side to side with a broad river of lava hundreds of feet deep. We saw the rhyolite from the road between Tower Falls and Cooke City—a high shelf rising on either side of the Lamar Valley, downstream from Buffalo Ranch.

## Late Lava Flows

Even after the rhyolite sea had congealed and cracked, black lava occasionally flowed and spread across the lowlands of Yellowstone. Some of these flows remain, resting on the rhyolite. You remember Oxbow Creek on the road between Mammoth Hot Springs and Tower Falls. About a quarter mile west of the creek, black lava rests on rhyolite in a road cut. The rhyolite of Gardiner Canyon's east wall upholds more basalt flows. You will recall that under the rhyolite lie other layers of lava, poured out before the sea of rhyolite submerged them.

## Fragmentation of the Plateau

Now that the lava had stopped flowing, Yellowstone began to split. Like a balloon, slowly emptying of gas and collapsing, the plateau settled. Earthquake paroxysms jarred the surface; segments dropped into the depths. Sometimes those segments lay flat, but more often the block slid into place in such a way that one end was tilted upward. It was as if a giant, sleeping below the surface of Yellowstone, stirred in his bed and rumpled the blanket of lava which covered him. Probably the wide, irregular basin of Yellowstone Lake—the largest lake at its altitude in the country—was the result of one of the giant's yawns. And after the basin was formed, streams poured into the trap from all the mountains now mirrored in the lake. The waters rose and overflowed to the north. They wandered in a devious course over the broken surface, into the valley of the Ancestral Yellowstone River. We know that river now simply as the Yellowstone.

The plateau settled in other places too, and other streams

rushed in to fill the depressions. Heart Lake and Lewis Lake and the broad basin of Shoshone filled while Yellowstone River was finding its way north among the broken plateau blocks. Some of the other basins, which may or may not have been flooded later, became partially filled with sediment from the surrounding plateau, and now their floors are level. Virginia Meadows and Gibbon Meadows and the level expanses of the geyser basins are sedimentary plains, but beneath the sediments the rock floor may be irregular. This Central Plateau where we stand was never disturbed by the giant's dreams. And Madison Plateau still stands at its original level.

At Madison Junction, and southward along the west bank of the Firehole River, runs a precipitous scarp, or cliff. Perhaps that scarp was formed when the Upper and Lower Geyser basins sank below the level of Madison Plateau.

Madison Canyon, a gateway carved through the plateau which separates the Snake River Plains from the geyser regions, is the western entrance to the Park. We shall see it soon. Once the Snake River Plains were on a level with Yellowstone Plateau, but the giant stirred, and the flatlands to the west subsided.

But basins were not the only evidence of the giant's restlessness. Purple Mountain, which rises northwest of us at Madison Junction, and whose western slopes disappear under the Snake River Plains, is a tilted block of the plateau. Like a huge seesaw, its eastern edge was flung high into the air and its western edge was thrown down and buried. Flat Mountain, which rises above the western shore of Yellowstone Lake, was tilted in the same way.

The Washburn Mountains had been tall before. They had risen along earth fractures long before the lava flood submerged their feet. And when the lava ceased to flow, the mountains stretched and grew hundreds of feet higher, and carried with them part of the plateau. Today we can see this broad bench jutting out along the southern segment of the range.

By courtesy of Haynes, Inc.

**RIVERSIDE GEYSER**

Even today the giant stirs restlessly, and now and then the crust of Yellowstone still quivers. The plateau surface is checkered with fractures, new and old.

### GEYSERS AND HOT SPRINGS

We have seen the geysers—fountains that stand out like white plumes when the sky is blue, or merge with the clouds when the sky is grey. They are the chimneys of Yellowstone, an integral part of this land of mountains and canyons. There are mountains and canyons in other lands, but few geysers outside of Yellowstone. Why do we find so many of them here?

Far below the plateau surface lies a mass of steaming rock. Geologists call it a batholith. From this rocky caldron come great quantities of highly heated vapors, and these vapors struggle to reach the surface. The upward journey is at first very difficult and slow, for the vapors must filter through infinitesimal crevices in the rock or soak their way upward like water through blotting paper. Shortly, however, the rising vapors meet the great fractures which formed when the plateau had broken up. The fractures penetrate deep into the crust, through the layers of volcanic rock which were spewed out by the Absaroka volcanoes, and even through those ancient sediments spread in layers when Yellowstone was born. Along these fractures the hot vapors rise rapidly, and their volume increases, for in their upward journey they convert to steam much of the water which once fell as rain on the surface of the ground but which later seeped into the rocks below. Finally the steaming vapors reach the level of the rock floor of the geyser basins, but they are still separated from the surface by the sediment which fills the basins. The vapors rise through these sediments in channels of their own making, and if they meet no obstruction, come out at the surface in vents called fumaroles. Sometimes the fumaroles are deadly, as in Death Gulch in the north part of the Park. Here, carbon dioxide escapes from below, and because it is heavier than air, lies along the ground—

a choking fog without color or odor, smothering every unsuspecting creature that comes within reach.

Usually, however, there is so much water in the rocks near the surface that a great part of the highly heated vapors are prevented from escaping into the air. They mix with the obstructing waters, heat them to high temperatures, causing them to rise, and geysers and hot springs result.

Wide channels allow the hot waters to rise freely to the surface, where the waters lose their intense heat and sink back to make room for more rising hot water. The circulation is free, then; no violent expulsion of water results. The miracle-tinted hot springs lie calm.

But there are longer and narrower channels, where the hot waters are choked and prevented from rising to the surface and losing their heat quietly. Nor can these waters boil readily, for boiling, which changes liquid to gas—as water into steam—can work only when there is room for expansion. Pressure may prevent expansion. The greater the pressure, the higher the temperature of the water must rise—higher than its normal boiling point—before the water can boil. But water will boil when its temperature has been raised so high that the pressure of its own vapors overcomes the pressure from above.

Now imagine a long, narrow tube, filled with water and heated from below. The water will be warmest at the bottom. Yet the lower the water in the tube, the higher its temperature must be for boiling—the weight of the column of water above presses down and must be overcome by more intense heat. We continue to heat the water, and inevitably every particle of water in the tube will at some time reach its own particular boiling point. Then if we lessen the weight of the water above, the whole column will boil at once. And in a geyser the surface water will finally be heated from below to its boiling point. It will boil and spurt a little. The pressure is lessened for a moment. Then the column streaks up in a cloud of steam and vapor. High in the air, the water cools, sometimes to run

MORNING GLORY POOL

By courtesy of Haynes, Inc.

back into the tube to be heated again, together with new supplies of water. Some of the geysers shoot out more water than their tubes can hold, so we believe that there is a subterranean chamber which the geysers use as a storeroom for water. Every geyser that we see bursts out from a tube which pierces the earth, a tube within which temperature and pressure struggle endlessly. It is a long story; we see only the majestic conclusion.

On their way to the surface, the rising waters dissolve some of the substance of the rock through which they pass. Such is the origin of the travertine which mantles the surface at Mammoth Hot Springs. The waters rose from the depths, and brought with them lime, from subterranean limestone formations. They laid it down as travertine. Those giant terraces, multicolored, at Mammoth, are made of travertine, but most of the deposits of the geyser basins of Yellowstone are of geyserite—a flinty substance dissolved from masses of underground rhyolite. If the travertine and the geyserite are precipitated close to the center of eruption at the surface, they build a cone; but if they are laid down at a distance from the center, then they form a terrace, and the waters overflow the first terrace to build a new one below. The deposit may eventually resemble a steep flight of steps.

Hot waters and vapors are responsible for much of the beauty of Yellowstone. The Grand Canyon, red and green and white and gold, might have been only the drab grey of rhyolite. But the vapors touch the rock and tint it. Some of the brilliance that stripes the walls comes from microscopic plants; but even the plants owe their color to the hot water seeping from the canyon walls.

All these things happened in the Volcanic Age—the age that brought to adolescence the Absarokas and the Tetons, the Washburn Range and the Red Mountains. It was an age of white-hot lava and tempestuous volcanoes. It was an age when violent earthquakes shook the placid meadows, earthquakes

MAMMOTH HOT SPRINGS TERRACE

By courtesy of *Haynes, Inc.*

that heralded the birth of mountains and the breakup of the newly formed lava plateau.

## LIFE OF THE VOLCANIC AGE

While all these things were happening, what had become of the little mammals, heirs to the kingdom of dinosaurs? At first none of them was larger than a rat. Most of them spent the greater part of their lives in trees. They were few in number.

But they were warm-blooded, and they were adaptable. They multiplied in hordes. They availed themselves of everything the new earth had to offer, and some of them grew to be of greater size than the largest elephant. In far-off Asia there roamed a beast as huge as some of the dinosaurs which lived before his time, twenty-five feet from head to tail and half again as high. He was the greatest land mammal of all time. There were giant birds who could not fly—birds like our own ostrich, but up to ten feet high. They laid their eggs on the western plains; more of the eggs have been found in foreign places. When one was found on the island of Madagascar, off the southeast coast of Africa, a tale was written of Sinbad the Sailor and that giant bird, the Roc; we have read the story in *The Arabian Nights*. And there were other birds, with heads larger than that of a horse, who raced their long legs over Wyoming's plains.

Our modern animals descended from the ancient mammals. Beasts such as the camel and the horny rhinoceros sprang from little creatures of our West, creatures with neither humps nor horns. And our horse was once a little five-toed thing, as big as a fox terrier. Camel, rhinoceros, and horse lived together in North America for millions of years. Camel and horse lived on through the Great Ice Age, and continued to roam the western plains until a few thousand years ago. Our modern horse is not a direct descendant of that early American creature; Spanish explorers brought him back to the Americas. The rhinoceros disappeared from North America near the close of the Age of Volcanism, and his place was taken by the elephant.

Like the dinosaurs, mammals fought for existence among themselves. The weaker tribes vanished, the stronger survived. And even the strongest often met defeat at the hands of nature. For during the Age of Volcanism, when volcanoes raged, day and night in the regions surrounding Yellowstone were one. It was an age of death and horror, when volcanic ash befogged the air, and every struggle for breath drew dusty suffocation into the lungs. Hordes of creatures died in dust; and those who lived through the dust were rewarded by starvation, for their forage lay buried in ash. And so the creatures of the Age of Volcanism left their bones in mute testimony of their tragic fate—an ancient forecast of the burial of Pompeii. And these bones, pitiful remains of the creatures of the past, lie in beds of ash all across the Great Plains.

After heat came cold. Burning lava was exchanged for glaciers. The furnace became a sea of ice, and the Great Ice Age swept upon Yellowstone.

ANIMALS OF THE ICE AGE

## V. THE GREAT ICE AGE

TODAY glaciers cover six million square miles of the earth, and in places are thousands of feet thick. But during the Great Ice Age, twelve million square miles of land lay buried under ice.

The Age of Volcanism ended with an icy breath. The continents were rising from the sea. The heat-preserving elements of the world's atmosphere were being depleted. And perhaps the sun itself faltered in its task of warming the earth. Winters grew longer, summers grew shorter. The icy wastes of the Arctic and the Antarctic crept over the plains and mountains of the temperate zones.

A field of ice thousands of feet thick lay over the northern half of North America. From Puget Sound on the Pacific to Long Island Sound on the Atlantic the ice field spread, its southern margin following the line of the Missouri and Ohio rivers. Half of North America became a new Antarctica.

And the creatures of the continent passed through another age of adjustment. Ice and snow crept over the vegetation; wind whistled around the mountaintops and swept across the plains. Rivers and ponds were frozen over. Little was left to eat and drink, and the bitter cold invaded every once-sheltered spot. Some of the animals migrated south, but others grew luxuriant fur coats. We think now of the elephant and the rhinoceros as rough-skinned, tropical beasts, but when ice clutched the land, the rhinoceros—who roamed the plains of northern Europe—and the elephant were hairy brutes. Far south of the ice front roamed the saber-toothed tiger and the creatures who feared him.

In many parts of the world there lived races of prehistoric men, primitive cave dwellers. Some were squat and hairy creatures who thought only of the harsh struggle for existence. But there were others, tall and handsome, who portrayed on their cavern walls and on their implements of bone the whole hairy horde of the Ice Age—elephant, rhinoceros, and strange beasts that no longer roam the earth. And during the last of the glacial age, man came to America, for his arrowheads are found among the bones of some of these ancient creatures.

### Ice Comes to Yellowstone

Even when sullen clouds hang low over Yellowstone, we find it difficult to believe that ice ever buried these green meadows and dense forests. Heavy though these skies are, how could they ever have let fall snow in such quantity as to give birth to a sea of glaciers? Yet green grass and purple peaks do change; we can imagine winter on this road, snow cresting the trees and mantling mountains, and pretend that the Ice Age was not so very different. But how imagine that transformation at the end of the Age of Volcanism, from lava to icicles, from suffocation to freezing? The burning sea gave way to a frozen waste, but the transformation required millions of years.

It happened slowly. Yellowstone lay hundreds of miles south of the North American ice field. But the lofty ranges which hem in Yellowstone on every side were feeding grounds for rivers of ice. Our highest equatorial mountains today breed glaciers, and Yellowstone's mountains were high too.

A million years ago the first ice came. Before it came, there were no deep canyons. Yellowstone River was cutting its way into the lava plateau, but the canyon was only a deep valley.

As the wintry blasts from ice-laden peaks swept over the plain, Yellowstone became a desolate waste. Winters lengthened, summer rains turned to snow, and every mountain peak was blanketed in white. Snow filled the air like volcanic ash, and each new flurry pressed down upon the snow and ice below until the frozen mantle covering the ground spread outward. And so, like the ice which now spreads from Mt. Rainier's cap in Washington, tongues of ice licked slowly down the mountain valleys toward the floor of the Yellowstone Basin. These were the glaciers. Like mammoth rasps, with fragments of rock for teeth, the glaciers gnawed at their valleys. They ate away the interlocking spurs of rock around which the streams had once twisted, and the winding V-shaped valleys became straight U-shaped troughs. As the troughs widened, the divides between them narrowed; some of them were ground into jagged, knife-edged ridges. And so the broad uplands of Yellowstone's mountain ranges were ravaged as if by disease—pockmarked and pitted by great canyons which spread like festering sores. Only a few isolated plots remained from all the rolling uplands.

We are standing in one of the most spectacular of the glacial troughs. Soda Butte Creek follows our road into the Lamar Valley. Think of it as filled with ice—a swollen river 3,000 feet deep and more than a mile wide. There was no road here at the beginning of the Ice Age, nor other human handiwork, only snow and ice, and desolation through ages, while icy tentacles groped down from the crests of the Gallatins, the Absarokas, and the Beartooth Mountains until they reached the Yellowstone Basin. The ice dammed some streams and forced others to new paths. It created lakes and wide marshes. And it pressed the Yellowstone River back, until its mounting waters overtopped the canyon and merged with Yellowstone Lake far to the south. From the lake the waters escaped over the Continental Divide into the Snake River and the Pacific.

The glaciers were not content to freeze Yellowstone once. Like the seas of Yellowstone's youth, they advanced and retreated and advanced again. For tens of thousands of years they stayed high among the mountaintops. Three times, and perhaps a fourth, Yellowstone was made into a land of snow and ice. The details of those early invasions we do not know; their evidence lies buried under fresher deposits or has been eroded away. But there were four glacial invasions in other ice-covered parts of the world, and Yellowstone was probably no exception. And though we know so little about the first invasions of the ice, we can guess at what must have happened; there are records of the last invasion, and they are sufficient. The same mountains which bred glaciers in the latter part of the Ice Age spawned ice in the early part. The same valleys which were choked with glaciers in the last glacial epoch were overrun in the first. The paths of the glaciers across the plateau were dependent on the topography and were similar for each glacial advance. And so the story of the last invasion is the story of each of the earlier ones.

### Early Invasions

We have seen the Grand Canyon from Mt. Washburn and from the road to Tower Falls, but now, halfway down the side of the canyon at Red Rock, we see a part not in view before. We had once thought that the sun caused the gleams of red and gold on the canyon walls, but today's grey skies hang over

GRAND CANYON

the same magnificence. The trail is steep, and descends halfway to the rushing torrent below. Half a million years ago, the river was hundreds of feet above its present level. Part of the ancient valley where it flowed lies behind us. It is largely filled with sediment now; but its outlines are distinct from the opposite side of the canyon. When the river had its course there, the canyon had not been carved deep enough to form Lower Falls, but for all we know, other falls may have shot over the ledge from a higher level to the floor of the valley.

From far off in the north, glaciers ground their way down the valleys of the Snowy Range and the northern Absarokas. They piled up in the valleys of the Lamar and the lower Yellowstone rivers until the valley floors were half a mile deep with ice. And there the imperious Yellowstone River was beaten. The great rampart of ice blocked its escape from Yellowstone; frustrated and vigorless, it dropped its load of sediment, until the valley was filled almost to the brim. If we follow the Red Rock trail, we can see remnants of those ancient sediments in the ancient valley.

But the climate warmed, and the ice melted, and the river found itself meandering over the flat surface of the sediment, high above its buried channel. Valleys spread wider at the top, and the new valley floor was several times the width of the old one. So the river, swinging across the plain, seldom ran directly over its first channel, but lay either to one side or the other. And near us now, at Red Rock, the Yellowstone pushed against the east wall of the valley, struggling to cut into the earth again. Here to the east of Red Rock, it carved a new channel, and cut it below the level of the old valley floor. And for centuries the river labored and carried away most of the sediment which littered its valley elsewhere, and the stretch of canyon between the two great falls was partly eroded. All this happened more than half a million years ago.

But before the canyon was completed, another long glacial winter set in. Ice dammed the Yellowstone again, and the weary river carried sediments from far away only to drop them on its own floor. Some of these sediments lie under Canyon Lodge—a house built on rocks that have lain there perhaps a quarter of a million years. We can see these rocks along the canyon rim and along the main road to the hotel. And along the course of Yellowstone River, between the canyon and Yellowstone Lake, more sediments prove that the lake drained to the north even then. For if there had been no valley then, sediments could not have been left there. A quarter of a million years ago, and we can still see the sediments! We begin—only begin, though we have been watching the work of eons—to feel the immensity of the ages behind us.

Then the glaciers melted away a second time. And while the earth was free of ice, the Yellowstone River struggled at its endless task. It had filled its valley again with sediment; now it must work to clear its path. And before the glaciers came once more, it had not only cleared its course of sediments but had deepened its valley, until the floor stood only fifty feet above the floor we see.

## Last Glacial Epoch

Now for the last time this Yellowstone of forests, the sanctuary of wild animals, this wonderland of geysers and hot springs, became ice-laden and barren. Glaciers rode down from the crests of the northern Absarokas and the eastern Beartooth Mountains. They merged in a gigantic stream which for the first few millennia of its existence remained confined in the Lamar Valley, down which it slowly moved. This glacier we know as the Lamar Glacier. Think of the slowness, the deliberateness of this river of ice, traveling then from the valley of the Lamar into the valley of the Yellowstone River near Junction Butte and down that valley to the north entrance of the Park at Gardiner, Montana, there to meet another flow of ice that sidled in from the Gallatins. Slowly and deliberately still, this gigantic flow of ice pushed its frozen snout thirty miles north of Yellowstone. It was tremendous—so tremendous and so distinct that men have given it a name and called it the Yellowstone Glacier.

That was a slow growth, an evolution. The Lamar Glacier had grown to vastness before it met the river of ice from the Gallatins. It grew in length, and as the climate turned steadily colder and more ice flowed from the highlands, Lamar Glacier grew thicker too. Finally it began to overflow to the south, up Yellowstone Canyon. It grew, and forced the waters of Yellowstone River to form a lake. And still it grew, and the lake water rose higher and higher. The northern shore of the lake was pushed southward by the spreading ice, while the southern shore moved farther south as the lake level rose, so that the lake may not have varied greatly in size. Eventually the great barrier of ice formed a dam so high that the waters of the lake spilled over the canyon rim. They flooded the land at the canyon head, filled the basin of Hayden Valley, and spread to join the waters of Yellowstone Lake. Here, where we see a canyon and a river, stood a dam of ice and a growing sea. Yellowstone River once more spread its load of sediment over its floor, gradually filling the canyon, so that the Lamar Glacier, forcing its way south, found the canyon shielded and was unable to grind away the lacy pinnacles which embellish the walls.

Years of this slow, relentless advance. And then came the day when the Lamar Valley was filled to the brim with the growing bulk of the Lamar Glacier. Now the ice which came down from the northern mountains found its way to the south made easier. Over the surface of the ice which filled the valley it rode, over the crest of Specimen Ridge, southward across the lava plateau. When its days of growth were over, the glacier covered a vast part of the Yellowstone Basin, burying the northern lowlands of Yellowstone under thousands of feet of ice. From the head of the Lamar Valley to its terminus thirty miles north of the Park it stretched, an ice stream eighty miles long. Most of our glaciers today are as insignificant beside this titan of ice as our elephants beside the dinosaurs.

The Lamar Glacier was not the only titan. Down the valleys of the southern Absarokas reached other rivers of ice which collected in the upper part of Yellowstone Valley, south of the lake. There they formed another huge ice stream, the Upper Yellowstone Glacier. Swollen, slow, the new river pushed northward between precipitous walls to the basin of Yellowstone Lake, and lay down, lazily, over the basin floor.

And so, like the jaws of a colossal vise, the Lamar Glacier closed in from the north, and the Upper Yellowstone Glacier ground its way up from the south. They squeezed the huge lake which stretched from ice front to ice front until its waters oozed through a channel in the Continental Divide to the west of Yellowstone Lake. The lake grew smaller; the fish were crowded into a smaller and smaller area, and the water was freezing. At last they fled to the low waters of the coast. And still the ice marched on. The bold white front of the Lamar Glacier pushed its way to the north end of Hayden Valley. The expanded front of the Upper Yellowstone Glacier filled the basin of Yellowstone Lake. Only Hayden Valley and the stretch of Yellowstone River Valley as far south as the basin of the lake were covered by water, and that water was escaping through a channel at the west end of Hayden Valley into the headwaters of the Madison River. With the same deliberateness, the ice fronts crept together, and met, and the last remnants of the lake vanished.

Cold it had been, and it grew colder. Ice had flowed down, and it flowed now in mightier streams. The journeys of the two glaciers temporarily ended. They piled on their own heads the new ice that crept down from the burdened highlands. The Upper Yellowstone Glacier rose higher and higher on the Continental Divide—the watershed separating the drainage of the Atlantic and Pacific oceans—until in some places it overtopped the divide. Tentacles of ice began to creep through low passes into Pacific drainage and to meet in the valley of Snake River. There they formed a new ice stream, which moved slowly southward to the foot of the Tetons.

And the Lamar Glacier rose too, and flooded across Central Plateau into the western geyser lands. A lobe of ice stretched through Madison Canyon and out of the Park. At its prime, the Lamar Glacier stood more than half a mile thick in the north. It buried the eastern peaks of the Washburn Range almost to their summits; perhaps a tongue passed through Dunraven Pass on its way south. Near Mammoth Hot Springs, where ice from the Gallatins joined the Lamar Glacier, ice buried the flat top of Mt. Everts, which rises 2,000 feet above the river at its base. And the head of the canyon, where in Canyon Lodge we shall sit tonight and toast our feet in front of a fire, lay under a thousand feet of glacial ice. Everything that we can see now was ice—forests and lakes and rivers, plains and valleys and canyons. Only the crests of the mountains and the summits of some of the plateaus lifted above the flood—and even their crests were white with snow.

## Waning of the Ice

Centuries rolled by, and the nightmare of glacial winter began to pass. Year by year the bitter winds grew softer. Where for countless centuries there had been only the rustle of settling snow, rain fell, loud and unfamiliar. And back from the lowlands melted the long fingers of ice until once more only the mountain peaks stood out in white.

But the glaciers left part of themselves behind them. They had come down out of the mountains laden with rock debris, some of which they had scraped and gouged from the bottom and sides of the valleys, and some of which had slid or washed down on them from the slopes above. When the glaciers melted away, all this debris was dropped, leaving a telltale trail over the regions once covered by ice. Such a trail of debris is known as ground moraine. The Beartooth Range is far to the north of us, and Inspiration Point is only a few miles down the canyon, yet the giant boulder that rests beside the road near the Point was carried there by ice from the peaks of the Beartooth Mountains.

The former position of the front of some of the ice streams is marked by a hummocky ridge consisting of an exceptionally thick deposit of rock debris. The reason for the ridge is simple. Imagine one of these tongues of ice, born among the lofty, snow-covered mountains, pushing its way to warmer levels below. When the glacier front has reached so low a level that the temperature is able to melt the ice as fast as it is urged forward, the glacier front—like a swimmer struggling against a strong current—will remain stationary. If the climate should turn warmer so that melting is more rapid, the glacier front would be melted back; if the climate were to turn colder so that the melting diminished, the glacier front would be pushed forward. Whether the front is stationary, retreating, or advancing, the ice itself always advances, for it is constantly urged down the valley by gravity. Since the ice is continually bringing new supplies of debris to the front of the glacier, where it is liberated by melting, the debris may pile up scores of feet high. Such a ridge, which outlines the position of an ice front, is known as a terminal moraine.

As the peak of the glacial winter passed in Yellowstone, the ice fronts, in their retreat, followed back over the paths they had come. Between the terminal moraine and the receding fronts of the long ice tentacles which stretched out beyond the Park, the gap grew wider and wider. One of the tongues of the Upper Yellowstone Glacier ended at the foot of the Tetons, where it dropped its debris. The debris became a dam that bound the water rushing from the melted ice. Now there stands an artificial dam, which has raised the level of the trapped waters still higher, and we call the water Jackson Lake; but thousands of years ago the lake was created by a Yellowstone glacier that rode down the valley of the Snake

River. There must have been other lakes—perhaps beyond the front of the ice stream which filtered through Madison Canyon, and beyond the terminus of the ice that crowded northward out of Yellowstone.

Through thousands of years the ice had come, through thousands of years it retreated. Time after time, in the cold of bitter winters, ice swept down from the mountaintops again, to revitalize the waning glaciers and recapture lost ground. But the climate was slowly growing warmer; the ice was defeated. Warm breezes opened the jaws of the Lamar and Upper Yellowstone glaciers, and once more lake waters rose between the ice fronts. Backward and backward crept the ice, till all of Hayden Valley lay submerged. For many years the lake lay in Hayden Valley. At that time the front of the Lamar Glacier stood near the head of the canyon; and the frozen snout of the Upper Yellowstone Glacier lay in Yellowstone Valley somewhere between Hayden Valley and the basin of Yellowstone Lake. Hayden Lake—where no lake is now—was hundreds of feet deep. Its rising waters escaped westward across Central Plateau into the headwaters of the Madison River.

Hayden Lake was not to stay long—not long in the millions of years of Yellowstone's history. Rivers from the melting ice and streams from the plateau surrounding the lake dropped their loads of sediment on its floor. After years of work, they filled the lake to the brim. And when the warming climate melted the ice fronts back again, Hayden Lake had disappeared, and in its place a broad upland plain stood between the lowlands of Yellowstone Valley to the north and south. Between that upland of Hayden Valley and the retreating front of Lamar Glacier lay a new lake, a lake whose level dropped with the waning ice. Once the Lamar Glacier had blustered up the canyon, driving a lake before it; now the lake waters crowded hard on the heels of the retreating ice.

Upper Yellowstone Glacier was dying too. The ice margin

of the glacier backed slowly southward. The waters of the melting ice flowed over the exposed strip of floor of the Yellowstone Lake Basin. The lake was then a crescent, formed around the margin of the glacier, and it overflowed into Pacific drainage through Outlet Channel. And the glacier front continued to move back, exposing the entire lake basin.

For centuries the new lake, Greater Yellowstone, stood 160 feet higher than our Yellowstone Lake. Its water spread over broad surfaces that lie above its level now. And just as they had done before and do today, streams left their burden of sediments along the lake shores. The shores broadened with every grain of sand, until, as time wore on, the sand closed in on the water. Like Hayden Lake before it, the lake seemed doomed.

But the game was not over yet. Yellowstone Lake, while the sand crept closer to smother it, was draining through the gap of Outlet Channel into Snake River and the Pacific. North of the lake rose the broad upland which the ancient sediments of Hayden Lake had founded. One of the streams that flowed northward across this upland was Restoration Creek. As it flowed, it gouged a deep valley in the plain. And the waters of Yellowstone Lake abandoned Outlet Channel for this northern pathway to the Atlantic.

Restoration Creek changed with the changing surface of Yellowstone. Lake waters were now escaping into the tiny stream, and Restoration Creek became a river, a new Yellow-

OUTLET CHANNEL

stone River. And it deepened its valley even more rapidly, causing the level of the great lake to sink. In a few years—as now we think of years—it had dropped a hundred feet. Around its shrunken shores there curved the broad shallow bench of the old shore line. The lake was probably the same clear blue, probably mirrored the trees and mountains in a glassy calm, just as it does today; but it still stood sixty feet higher than we see it. Perhaps its surface remained at this level because Restoration Creek, cutting through the sediments of old Hayden Lake, bit into the rocky wall of buried Yellowstone Valley. And the rock may have resisted erosion and held up the level of the lake. For years it remained that way, until the streams had built another shore terrace of sediment, a hundred feet below the older bank. Then the lake sank to its present level. So now all around the lake, especially there in the east, we can see the two broad benches—giant steps leading down to the blue waters below.

Upper Yellowstone Glacier was now a shrunken fragment of its former self. And the proud flow of ice called Lamar Glacier was wasting away in the north. Ice melted, and the streams that flowed between ice and valley sides gouged channels in the rocky slopes. We saw some of the channels on the northern slopes of Yellowstone Valley, across from the Ranger Station at Tower Falls—abandoned, deep-cut gorges on the faces of the hills.

And then for scores of years, a lobe of the Lamar Glacier, like Horatio at the bridge, held the pass between Specimen Ridge and the lower slopes of Prospect Peak. Feebler and feebler it grew; but for centuries it thrust back the surge of the Yellowstone River. And it formed a dam which held in a lake eight miles long and 500 feet deep. The sullen, ice-swollen waters of Retreat Lake drowned Yellowstone Canyon and flooded the valleys of Antelope and Tower creeks. They overflowed to the northwest, along the margin of the ice, and rushed across the barren top of Overhanging Cliff.

The old gravel that now lies on top of the cliff is similar to the gravel in the bed of Yellowstone River, 600 feet below.

Upstream from Retreat Lake, the river labored to free the canyon of the binding sediments of the Ice Age. As the Mississippi carries its burden of rock and gravel into the Gulf of Mexico, so the Yellowstone spread the debris over the floor of Retreat Lake. More and more sediment, higher and higher the floor, shallower and shallower the lake. Where ice-splotched waters had buried Yellowstone Canyon and beat against the front of Lamar Glacier, there now lay a flat sedimentary plain. Curving over the plain flowed Yellowstone River—wide and clear, meandering across the level surface until it swept over the summit of Overhanging Cliff. For years it continued thus, until the ice which had bound the river melted. Then the new lower levels of ice on the slopes opened new paths to the river; it abandoned the high crest of the cliff, and passed between the face of the cliff and the wasting ice.

The river was unchecked now. Freed of the irksome yoke imposed by the confining ice, the river gave up its listless wanderings and became as we see it now. It rushed down from the surface of the plain, carrying away with it vast quantities of the sediment of Retreat Lake. The deeper the river cut, the higher above it stood the remaining sediments. Looking upstream from the platform at Tower Falls, we can see low spurs in the distance, reaching out from the cliffs on the far side. These spurs are carved out of the ancient sediments of Retreat Lake, and their surface shows us where the lake level once stood. Some of these sediments are exposed in the road cuts at Tower Falls and cover the upper layer of black basalt across the canyon.

We know, too, why Tower Falls is here, plunging down among the sentinel rocks. When Yellowstone River deepened its valley, Tower Creek deepened its own smaller channel. Hard rock met and resisted its force, and the stream plunged across and downstream, to carve the weaker rock there into a less stubborn channel. We stand near those resistant rocks.

Ten thousand years ago, the last lobe of ice melted away from the Yellowstone lowlands. Only ten thousand years ago—little time for convalescence from that icy ordeal. Debris that dropped from the ice still mantles the rocks over most of Yellowstone. Many of the old sediments which were laid down in Yellowstone Lake, and Hayden Valley, and the Grand Canyon, still remain. On the highest peaks the struggle between heat and cold still rages. For ten thousand years the tattered glaciers have battled. A remnant of that vanquished host is Grasshopper Glacier, within which millions of grasshoppers, blown upon the ice and buried under successive snows, lie frozen. The Great Ice Age did its work, left its monuments, and departed.

LOWER FALLS

## VI. FUTURE OF YELLOWSTONE

WE HAVE imagined Yellowstone from the earliest days of its history—from the time of Cascadia and Appalachia, when the whole of Yellowstone was nothing but a barren plain. And we pretended to watch the seas sweep in, Arctic and Pacific together, drowning the plain and receding and flooding inland again. Here, on this densely wooded plateau, we stood and thought of those first years, when the land was beginning to bear fruit. On the road to Tower Falls we saw the mountains that bind Yellowstone in a rim of craggy summits, glimpsed the brilliant patches of red sediment on the flanks of the northern hills, and tried to imagine the days when Yellowstone was mountainless and parched. The Sundance Sea and its monsters; dinosaurs and flying dragons; the volcanic birth of these grey-purple Absarokas to the east; earthquakes that heralded the grandest mountains, the Tetons that barricade Yellowstone to the southwest—we thought of all those things. We found at the foot of Crescent Hill the plant-encrusted rocks from the earliest lava flows; we saw Overhanging Cliff and the place beneath it where geologists once found the bones of an ancient horse.

Geysers are spouting against the sky now; the tubes under the earth are brimming with water boiled deep in subterranean caldrons. We have imagined the time when lava scalded the earth, before these geysers began; when dust choked the air, and Yellowstone was the home of only a few wild beasts, who feared their carnivorous neighbors more than smoke and flame. Now the deer and buffalo live here in contentment.

We thought back to the Ice Age which froze the plain and changed the surface of Yellowstone again. The Lamar Glacier and the Upper Yellowstone Glacier met like the jaws of a vise to squeeze Yellowstone Lake between them; little Restoration Creek grew to be the Yellowstone River; and the Grand Canyon of the Yellowstone was forged into its castled grandeur. And then, ten thousand years ago, the ice melted away. Centuries later, Indians came to hunt in the forests and fish in the streams. Hunters and trappers struggled through the wilderness. And John Colter, because of an adventurous spirit and a wounded leg, discovered Yellowstone Park.

This is our last hour in Yellowstone. We have read the past; shall we dare foretell the future? If we might come back to this exact spot a few million years from now, what should we see?

Nothing of this. Nothing of the geysers that spout into the sky. Nothing of the springs with their rainbow tints, welling up from the tubes below. There will be no store of heat to shoot boiling water and steam higher than the pines. For the rocks below the pines will cool. Before this happens, however, earthquakes will have closed the vents of some geysers and opened new ones, as they are doing even now. Who knows how soon Old Faithful will cease to be?

There will be no waterfalls—Upper and Lower Falls, Virginia Cascades, Tower Falls. Yellowstone Lake, Heart Lake, and double-bowled Shoshone will be gone. Through the ages, the old resistant rock that holds up Lower Falls will disappear. The sediment-laden waters, swirling across its rocky bar, will wear it down. And between the vanishing brink and Upper Falls, the canyon will grow deeper. Long after Lower Falls have made their last plunge to the river below, the broad belt of rock which holds Upper Falls will stand. Deeper than ever the canyon will grow; and the falls will be for a time a cataract greater than all these we see—a thing of magnificence, 400 to 500 feet high, leaping to the canyon floor.

But as the years come and go, that rock too will wear away, and the Yellowstone Valley, as far as Yellowstone Lake, will be cut down again. The level of the lake waters will slowly lower until, when the valley is eroded down a few more hundreds of feet, there will be no lake. Yellowstone River will usurp the abandoned lake floor. And all the lakes in Yellowstone will suffer the same fate; they will be drained as the outrushing waters cut their channels down.

Canyons will eat deeper into the heart of the lava plateau. And the canyons will dig and spread, until they lie one against the other, and Yellowstone becomes a raw, jagged wilderness. And the wilderness streams will gnaw everlastingly at the hills—Mt. Everts, Elephant Back, Purple Mountain—and carve them down to their roots. Yellowstone will once again become a barren and fruitless plain. And only the twisted rocks laid bare at the surface will tell some geologist of the future what has happened here.

All this will happen in a few million years. Unless there is another Ice Age. Unless new mountains grow from the roots of the old. Unless volcanoes bring on another age of fire.

Now we must leave Yellowstone and go back to our homes. We must accustom ourselves to thinking again in terms of years rather than eons. The skyscrapers and bridges and trains that man has built and the swift metal birds that he has made to fly over them will thrill us as before. Yellowstone will seem far away. But we shall remember these days in Yellowstone all our lives. We have shared intimately in the fascinating drama of the planet which carries us through space; we have gained a wholesome perspective from staring into the vast depths of time; and we have beheld in the grandeur of Yellowstone the living symbol of the magnificence of America.

# North America

During birth of Yellowstone

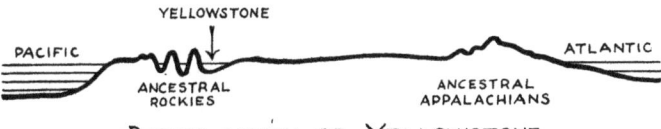

During youth of Yellowstone

From close of Yellowstone's youth to now

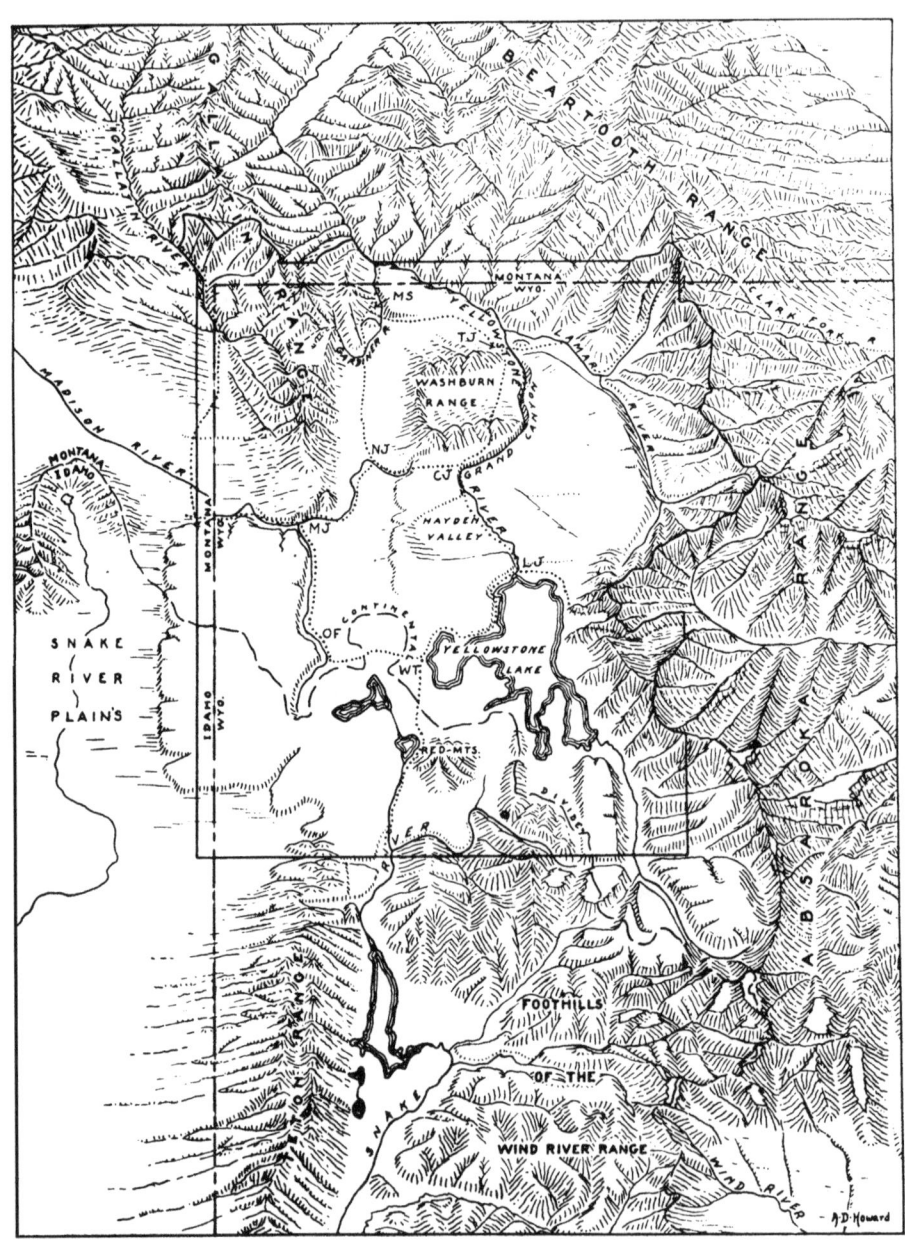

YELLOWSTONE PARK AND VICINITY. The roads in the Park are indicated by dotted lines and the principal junctions by letters. Mammoth Hot Springs (MS), Norris Junction (NJ), Madison Junction (MJ), Old Faithful (OF), West Thumb (WT), Lake Junction (LJ), Canyon Junction (CJ), Tower Junction (TJ). The area is 130 miles long and 100 miles wide.

Bei Fragen zur Produktsicherheit wenden Sie sich bitte an:
If you have any questions regarding product safety,
please contact:

Walter de Gruyter GmbH
Genthiner Straße 13
10785 Berlin
productsafety@degruyterbrill.com